discovermore

Learning Social Studies Skills

Reading Maps

Marie Harts

Britannica Educational Publishing
IN ASSOCIATION WITH
THE ROSEN PUBLISHING GROUP

Published in 2025 by The Rosen Publishing Group, Inc.
2544 Clinton Street, Buffalo, NY 14224

Editor: Shannon Harts
Book Design: Michael Flynn

Photo Credits: Cover Alexander Lukatskiy/Shutterstock.com; (series background) Dai Yim/Shutterstock.com; p. 5 (top) frantic00/Shutterstock.com; p. 5 (bottom) Roman Samborskyi/Shutterstock.com; p. 6 Mike Flippo/Shutterstock.com; p. 7 siriwat sriphojaroen/Shutterstock.com; p. 8 Ron Zmiri/Shutterstock.com; p. 9 Melinda Packer/Shutterstock.com; p. 10 K. Ramirez/Shutterstock.com; p. 11 Gorodenkoff/Shutterstock.com; p. 12 Sergey Mikhaylov/Shutterstock.com; p. 13 SevenMaps/Shutterstock.com; p. 14 Piotr Przyluski/Shutterstock.com; p. 15 Alexander Lukatskiy/Shutterstock.com; p. 16 sjgrafics/Shutterstock.com; p. 17 Jim Pruitt/Shutterstock.com; p. 18 Paul Brady Photography/Shutterstock.com; p. 19 Libin Jose/Shutterstock.com; p. 20 Daniel Jenny/Shutterstock.com; p. 21 O.Kemppainen/Shutterstock.com; p. 22 Wangkun Jia/Shutterstock.com; p. 23 Claudio Divizia/Shutterstock.com; p. 24 Siberian Art/Shutterstock.com; p. 25 Kollawat Somsri/Shutterstock.com; p. 27 (top) AngieYeoh/Shutterstock.com; p. 27 (bottom) wee dezign/Shutterstock.com; p. 28 Svetliy/Shutterstock.com; p. 29 Phovoir/Shutterstock.com.

Cataloging-in-Publication Data

Names: Harts, Marie.
Title: Reading maps / Marie Harts.
Description: New York : Britannica Educational Publishing, in Association with Rosen Educational Services, 2025. | Series: Discover more: learning social studies skills | Includes glossary and index.
Identifiers: ISBN 9781641901567 (library bound) | ISBN 9781641901550 (pbk) | ISBN 9781641901574 (ebook)
Subjects: LCSH: Map reading--Juvenile literature. | Maps--Juvenile literature.
Classification: LCC GA130.H347 2025 | DDC 912.01′4--dc23

Manufactured in the United States of America

Some of the images in this book illustrate individuals who are models. The depictions do not imply actual situations or events.

CPSIA Compliance Information: Batch #CSBRIT25. For further information contact Rosen Publishing at 1-800-237-9932.

Contents

Defining Maps

If you want to travel the world one day, there is one tool that you will want to make sure you have with you. This same tool can help you explore your school and neighborhood. This tool is a map! Imagine how hard it would be to travel to a new place without a map.

A map is a drawing or visual of all or part of Earth's surface. Most maps are drawn or shown on a flat surface. A map displayed on a round surface is called a globe. Globes give the same kinds of information that flat maps do. Because Earth is almost round, a globe represents it best. A globe shows Earth as it looks when seen from space. Maps may show visible features, such as rivers, forests, buildings, and roads. They also may show things that cannot be seen, such as where one state is separated from the states around it. Special maps show other invisible things, like weather patterns. Although maps come in many different forms, every map can provide useful information about the world around you!

If you are traveling in the United States and need a paper map, you can often find one for free at a regional, state, or city visitor center.

compare and contrast

What do you think are some similarities and differences between paper maps and globes?

Globes are scale models of Earth. This means they show accurate information regarding distances and the shapes of objects. However, most globes aren't a great size for taking from place to place when traveling!

Common Types of Maps

Political maps and topographic maps are two types of maps that are commonly used. Topographic maps show the location and shape of features on Earth's surface. These features often are natural things, such as mountains and lakes. Topographic maps can also show things that humans have built, such as roads and railroads.

Political maps show the boundaries of countries, states, provinces, counties, and cities. Boundaries are the unseen lines that separate places run by different governments. Some visible features, like rivers and mountain ranges, also can act as boundaries. These features physically separate one place from another. Most maps combine features of both topographic and political maps.

Topographic maps usually have contour lines that show the elevation, or different heights, of areas.

Maps that show subway lines can be helpful for those who live in and visit cities.

Some special maps show things that are not on Earth's surface. For example, star maps show the stars in all or part of the sky. Maps can also show what is underground, such as **infrastructure** for subways and underground railways.

WORD WISE

INFRASTRUCTURE IS THE BASIC PHYSICAL STRUCTURES, SUCH AS ROADS AND TRANSPORTATION SYSTEMS, NEEDED BY AN ORGANIZATION, REGION, OR COUNTRY.

The Compass Rose

A helpful feature on many maps is a compass rose. This symbol got its name from its shape, which looks like a flower. Arrows or pointers stick out from the center, like petals. These arrows show the cardinal directions—north, east, south, and west.

Some maps have compass roses showing northeast, southeast, southwest, and northwest. These points are called the intercardinal directions. This name comes from the prefix "inter," meaning "between." It refers to a direction between two of the cardinal directions. Other maps lack a compass rose. Instead, they may have just an arrow pointing north.

The compass rose was invented around the 1300s. It was first used to show wind direction and was called a wind rose.

A weather vane, like a compass, shows the cardinal directions north, east, south, and west. It also shows the direction of the wind!

Maps often are printed so that north is at the top, but that is not always the case. Therefore, the compass rose helps map users to figure out which way is which. Determining how directions on a map relate to directions on the ground is called orientating.

Consider This

What are some reasons it might have been important to know wind direction back when the compass rose was invented in the 1300s? Why might it be important to know about wind direction today?

Learn About Map Legends

You will often find multiple different symbols on a map, from stars to triangles and even tiny trees! Many maps have a legend, also called a key, where you can learn what each symbol means. It can also help you understand other visual aids, including colors.

Most of the time, a map legend can be found in a separate box or on the side of a map. It often shows symbols and colors on the left and definitions on the right. For example, the legend on a map of Washington, D.C., might have dots and squares next to labels. The labels could explain that dots mean train stops and squares mean places to visit.

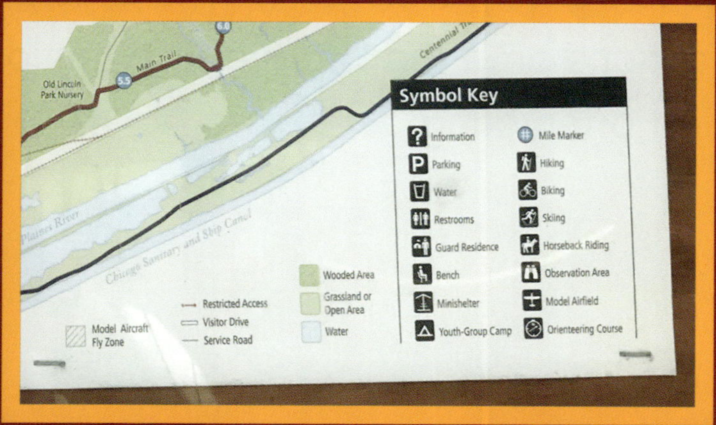

Many maps of parks have legends that show what symbols of important places mean, including parking and bathrooms.

Weather maps often have lines with blue triangles showing low **air pressure** and lines with red half-circles showing high air pressure. A TV meteorologist acts like a map legend to help viewers understand what these symbols mean.

Different types of maps have different types of legends. The legend for a map of a park might have symbols for paths and picnic areas. A weather map legend often shows that blue symbols mean cold weather and red symbols mean hot weather.

WORD WISE

AIR PRESSURE IS THE PRESSURE CREATED BY THE WEIGHT OF THE AIR THAT PRESSES AGAINST EVERYTHING IT TOUCHES.

Simplifying with Symbols

Many early maps included drawings to show features such as mountains and castles. The drawings were large so they could be seen clearly. Such artwork was nice to look at and easy to understand. However, it produced maps that were cluttered and not very detailed. Over time, mapmakers made maps less fancy. They replaced pictures with symbols.

Multiple maps from **medieval** Europe include drawings of sea monsters that many people really believed lived in the seas at the time!

Many maps have nearly universal symbols. These include an "H" for a hospital, a cross for first aid, a fork and knife for food, and a bed for lodging areas.

On modern maps, symbols are made of simple shapes and lines. Each symbol represents a particular thing. A single black line might stand for a road. Double black lines might stand for a highway. Symbols for natural features have unique shapes and patterns. Mountains are sometimes shown as triangles or as upside-down "V" shapes. Forest symbols may look like groups of tiny trees or green patches. A scattering of dots is a common symbol for deserts. Good symbols stand out from the background of the map. They should make it easy to understand a map's information from a quick glance.

WORD WISE

MEDIEVAL MEANS HAVING TO DO WITH THE MIDDLE AGES, A TIME IN EUROPEAN HISTORY FROM ABOUT 500 TO 1500.

Cool Colors

Features on maps are usually marked by different colors. Maps often use the colors we usually think of when we imagine natural features. For example, what color reminds you of water? You probably thought of blue. On a map, rivers, lakes, oceans, and other bodies of water are usually shown in blue. Light brown or tan often shows where deserts are.

On many maps showing bodies of water, a darker blue color shows deeper areas and lighter blue is used for shallower areas.

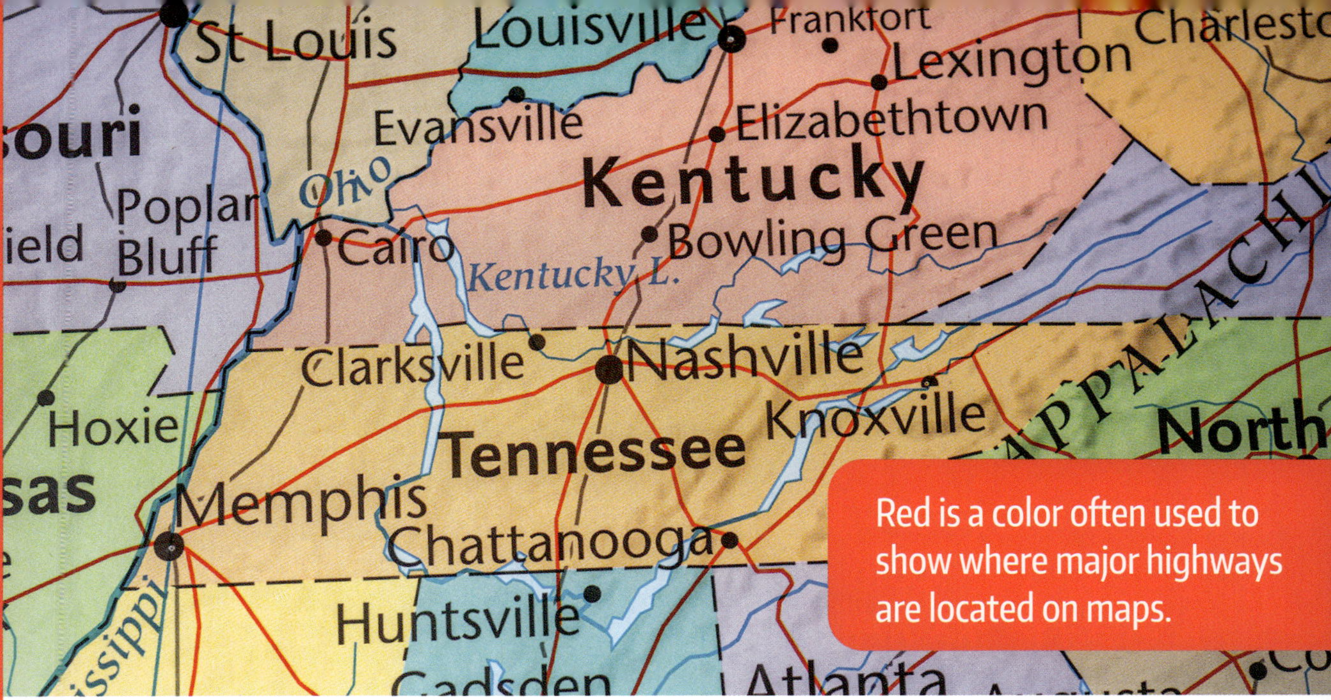

Red is a color often used to show where major highways are located on maps.

Colors also can be used to show invisible or unseen qualities. A population map might use darker colors to mark where more people live. Lighter colors would show where fewer people live. Maps that show countries, states, or other political features often use bright colors and bold lines. This **design** can make places on maps clearer to see.

Consider This

What do you think the color green is often used to show on maps?

WORD WISE

A DESIGN IS THE PATTERN OR SHAPE OF SOMETHING. IT IS ALSO TO CREATE THE PATTERN OR SHAPE OF SOMETHING.

Distances on a Map

If you are looking at a map of a park, the distance from the parking lot to the bathroom is much smaller on the map than it is in real life! Most maps include something called a scale to show how much smaller the map is. The scale shows how distances on the map are related to the actual distances.

Circumference is the distance around Earth, a globe, or any other sphere.

Consider This

Globes often have fewer distortions, or changes that make something less true or accurate, than flat maps. However, they can still have some distortions or deletions based on cost or a globemaker's beliefs. What do you think these could be?

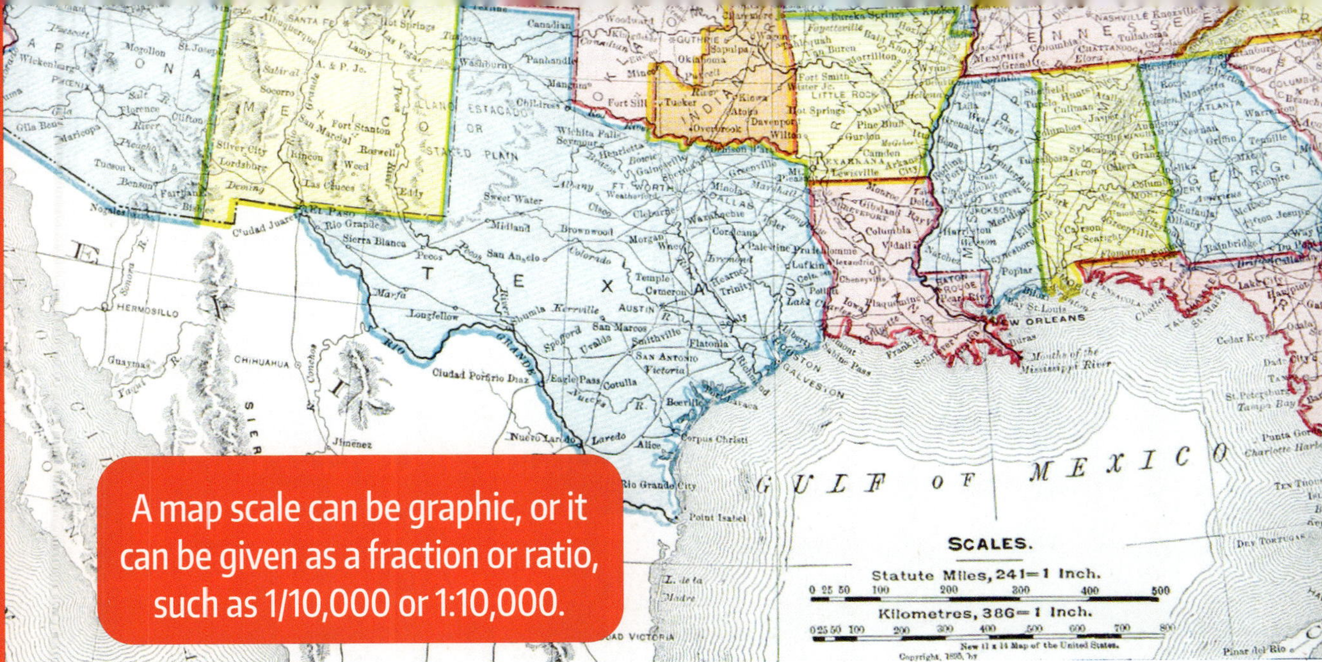

A map scale can be graphic, or it can be given as a fraction or ratio, such as 1/10,000 or 1:10,000.

Map scale can be shown in different ways. The simplest way is a graphic scale. A graphic scale is a ruled line or bar that is usually marked off in miles or kilometers. A graphic scale can be used to measure distances on the map. Some maps have a verbal scale. A verbal scale uses words, such as "one inch equals 60 miles."

Compared to most maps, globes show a much larger area—the entire surface of Earth. The circumference of many globes is 3.14 feet (0.96 meter). The circumference of Earth is 24,902 miles (40,075 km). However, globes are said to represent distances more accurately than flat maps because they are spherical, or ball-shaped.

Showing Cities and Towns

When driving across the country, it can be fun to spot a sign telling you that you've entered a new state. Perhaps the sign was the only thing that showed the state line. Some places are separated by rivers or other natural features. Others are separated by walls or fences. But many places are separated only by imaginary lines.

A capital is the city where a region's government is found. The capital of New York State is the city of Albany, New York. It is where you can find this building where the state's government meets.

The names of countries are often spelled in capital letters on political maps.

Political maps show these invisible borders. They show where one country or state is separated from its neighbors. They also may show other government boundaries, such as lines that divide states into counties or districts. Besides borders, political maps give us detailed information about a place, including its cities and towns. Dots show where cities are located. Dots for major cities are often bigger than those for smaller cities. A capital city is often marked by a star.

Consider This

Bodies of water are a feature of nature, but they are often shown on political maps. Why do you think this is?

Showing Earth's Features

Political maps often show invisible state and country borders. Topographic maps, on the other hand, show the visible features of Earth. These features include natural things you can see, like rivers, lakes, mountains, hills, deserts, forests, and other plant life. They also often include visible features that are man-made, like roads, railroads, and walls.

Topographic map contour lines can allow you to see a three-dimensional (3D) representation of a map's flat surface! Three-dimensional means having or seeming to have depth, length, and width.

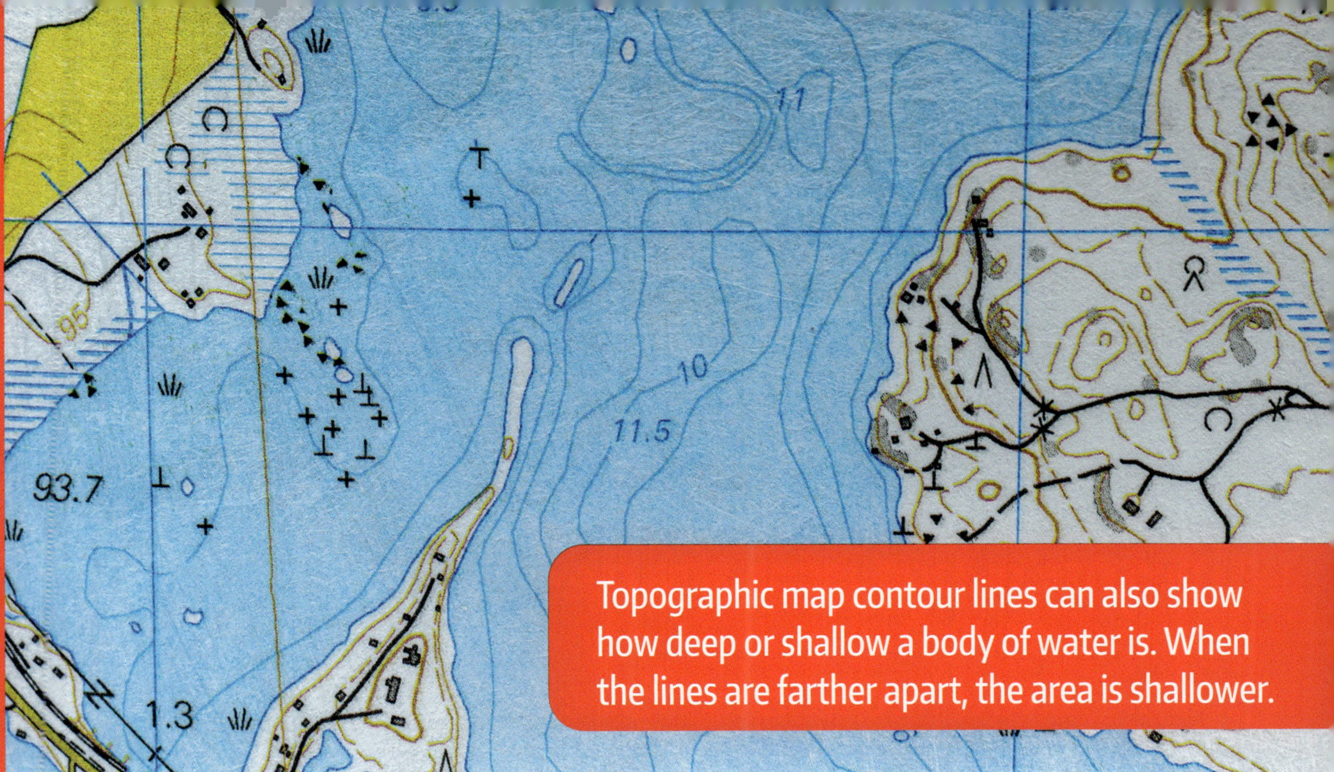

Topographic map contour lines can also show how deep or shallow a body of water is. When the lines are farther apart, the area is shallower.

Areas of Earth that are underwater also have natural shapes that can be mapped. Some topographic maps and globes show physical features found under the oceans and other bodies of water. This type of topographic map is useful to scientists and sailors. An important feature of most topographic maps is contour lines. These lines connect points of the same elevation. The closer together these lines are, the steeper the **terrain**.

WORD WISE

TERRAIN IS A WORD USED TO DESCRIBE PHYSICAL FEATURES OF A PIECE OF LAND.

Transportation Maps

Whether you are visiting or living in an area, it's important to know how to get from place to place! At one time or another, almost everyone has to drive or take public transportation somewhere. The main purpose of road maps is to provide drivers and other travelers with information on how to get somewhere using streets, roads, and highways. Different colors and designs in a road map's legend usually mark different kinds of motorways. Some may be toll roads that require a toll, or fee, to drive on them.

Most subway maps are found on walls or signs at subway stations.

On the New York City subway map, different subways are identified by certain numbers or letters.

Many cities, large towns, and counties have their own public transportation systems. Systems in very large cities might have dozens of different bus, train, or tramway lines. Maps of these systems often are color coded to help passengers make choices about where to get on and off a bus or train. These maps also show where different bus and train lines meet so that riders can hop onto a different line to get where they are going!

Consider This

When using public transportation, there are some things to keep in mind. These include making sure to buy tickets and to check a schedule to know when a train or bus arrives. Public transport can be crowded. What manners do you think are important to follow?

Pinpointing Places on Maps

Most maps have certain features that can help you find a particular place—one example is latitude and longitude. Latitude and longitude are sets of imaginary lines that circle Earth. Latitude runs east and west. Longitude runs north and south. Any place can be found on a map using its measurements of both latitude and longitude.

Latitude and longitude are both measured in degrees (°). Lines of latitude go from 0° at the **equator** to 90° north (N) at the North Pole and 90° south (S) at the South Pole.

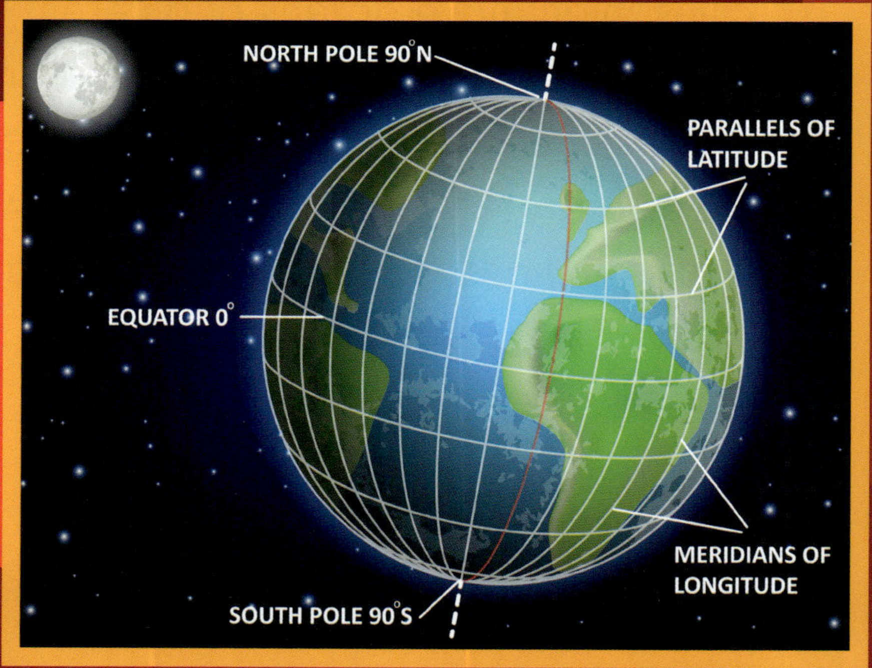

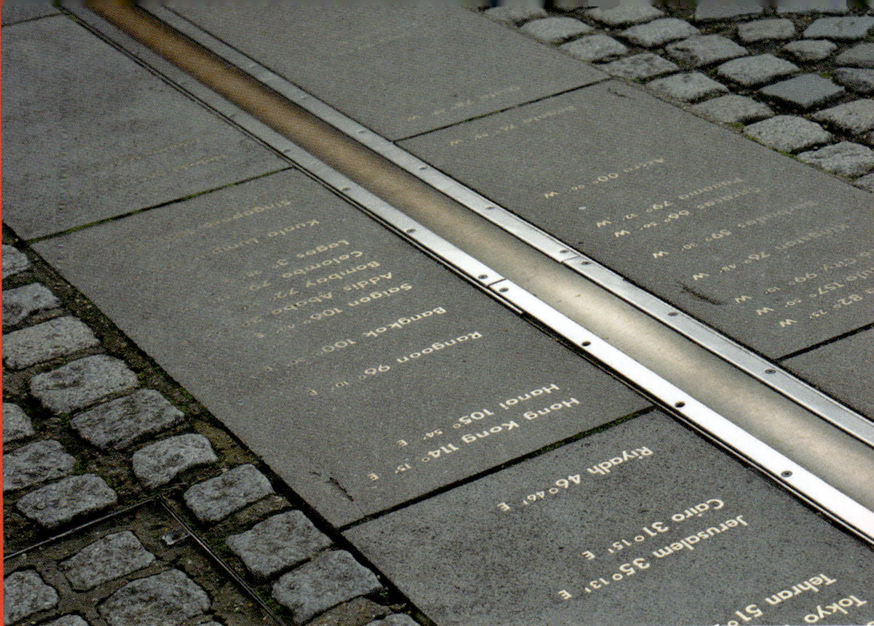

Longitude is a measurement east or west of the prime meridian, a north-south line that passes through both the North and South poles. Although the line is imaginary, it is marked on the ground at the Royal Observatory in Greenwich, London.

Some maps are divided into sections like squares on a checkerboard. Rows across may be lettered "A," "B," "C," and so on. Columns that go up and down may be numbered "1," "2," "3," and so on. These maps often have an index that tells where to find a place by using the letters and numbers. If the index says a place is located at "B 2," for example, the place can be found where row "B" and column "2" cross. Road maps often use this system. People call the latitude and longitude of a place's location or the letter and number of a place's location its coordinates.

WORD WISE

THE EQUATOR IS AN IMAGINARY LINE AROUND EARTH THAT IS THE SAME DISTANCE FROM THE NORTH AND SOUTH POLES.

Maps Meet Today's Tech

Although paper maps are still quite popular, many people also use maps on smartphones, laptops, tablets, and computers. These are called digital maps and they often require an internet connection to access. Online digital maps are useful in many ways.

The information on many digital maps comes from satellites orbiting Earth. Satellites take detailed pictures of streets and neighborhoods. These pictures are used in many digital maps. Some digital maps also use pictures taken by cars that drive around on Earth's surface. Such pictures are useful because they show you exactly what a place looks like.

One of the most popular uses of online maps is to get directions. When you enter an address or street intersection, the map program tells you how to get there. The map might give you different directions based on whether you select that you are biking, walking, driving, or taking a train or bus.

Google Maps was the most downloaded map app for navigation in the United States in 2022. Navigation is the science of plotting and following a path from one place to another.

compare and contrast

What are some similarities and differences between paper maps and digital maps? When do you think it's better to have a digital map, and when are paper maps better?

While navigating with digital apps like Google Maps can be useful, studies have shown that they can harm people's sense of direction and it's better to use paper maps.

Become a Cartographer!

A cartographer is someone who creates maps. Making your own simple map can help you learn about them! You can start by picking a few places in your town or city that are important. Include your home, school, and other places that matter to you.

You could turn a map you make of your backyard or home into a treasure map! Simply hide something that could be considered "treasure" and use your map as a guide to find it.

Then make a legend. Draw some symbols or simple pictures that represent buildings, parks, and other destinations. Then bike, walk, or have a family member drive you to these places to figure out how to get to them. Draw the roads you took on your map. Next, using the symbols you made, mark your destinations on the map. For a challenge, add cardinal directions in a compass rose to your map. Finally, share the map you drew with someone. This experience might lead you to one day become a **professional** cartographer who creates maps to help with urban and regional planning!

Making a map can help you with spatial thinking. This important kind of thinking allows you to picture in your mind where objects and places are in relation to each other.

WORD WISE

PROFESSIONAL MEANS HAVING TO DO WITH A JOB.

Glossary

cluttered Messy or confusing to look at.
compass A device with a magnetic needle that points north.
digital Characterized by electronic and computerized technology.
fraction A number that indicates one number is divided by another number, meaning it is only part of a whole.
information Knowledge obtained from study or observation.
invisible Unable to be seen.
meteorologist Someone who studies weather, climate, and the atmosphere.
province A large division of a country that has its own government.
ratio The relationship between the numbers, size, or amount of two things. It is often shown by two numbers separated by colons.
regional Having to do with a part of the world or of a country that is different or separate from other parts in some manner.
satellite An object that circles Earth in order to collect and send information or to aid in communication.
sphere A three-dimensional object that is completely round.
symbol A picture, shape, or object that stands for something else.
tramway A form of public street transportation that runs on rails and is often powered by overhead electric cables.
universal Occurring everywhere.
urban Relating to cities and the people who inhabit them.

For More Information

Books

French, Vivian. *From Here to There: A First Book of Maps.* Somerville, MA: Candlewick Press, 2023.

Mizielińska, Aleksandra, and Daniel Mizieliński. *Maps.* Somerville, MA: Big Picture Press, 2020.

Websites

Map and Globe

kids.britannica.com/kids/article/map-and-globe/353425

Learn more about maps and globes from Britannica Kids!

Nature Cat: The Map!

pbskids.org/video/nature-cat/2365604613

This fun animated video will help you learn about using a map to find features of an area.

Publisher's note to educators and parents: Our editors have carefully reviewed these websites to ensure that they are suitable for students. Many websites change frequently, however, and we cannot guarantee that a site's future contents will continue to meet our high standards of quality and educational value. Be advised that students should be closely supervised whenever they access the internet.

Index